Enduring Threads

Barbara Pyett

Enduring Threads

GP

Enduring Threads
ISBN 978 1 76109 425 5
Copyright © text Barbara Pyett 2022
Cover image: Barbara Pyett

First published 2022 by
GINNINDERRA PRESS
PO Box 3461 Port Adelaide 5015
www.ginninderrapress.com.au

Contents

Author's Note

I'd like to begin by acknowledging the Bunurong people, Traditional owners of the land in which we live today, and pay my respects to the Elders past and present. I extend that respect to all Aboriginal and Torres Strait Islander peoples.

This autobiographical sestina collection has been written in this form as a self-imposed constraint to restrict my verbosity. The sestina was invented by the twelfth-century mathematician Arnaud Daniel. The end words of the first stanza are repeated in a set formula throughout the poem. I have found this form of poetry thrilling and addictive.

A big thank you to all of the Sandybeach writers for their support, especially Darrelle Spenceley, who has helped this compilation come into being with her generous time and help as editor. Also I wish to thank Brenda and Stephen Matthews at Ginninderra Press for their patience and for initiating me into the intricacies of producing my first poetry book. Finally, I want to thank Christopher for his encouragement and for being my patient sounding board and soulmate.

Devonport, Tiagarra, Tasmania, 1949

Mother dressed, chose piebald shoes,
silk-stockinged feet slipped into heels,
powder, lipstick applied with care.
'Why do you put powder on your face?'
I'd ask. 'Wait until you're grown-up,
you'll do the same, no doubt.'

I watched with wonder and in doubt
that I would ever do the same. My brown shoes
polished bright, laced up
with rubber-soled flat heels,
red kilts alike, hands, knees and face
scrubbed clean, plaits pulled tight, no care

or thought of how that felt, red ribbons carefully tied with bows to match our kilts. Doubt
never crossed my mind, her love sincere, soft face,
sprayed with Poeme L'eau. Time to put those shoes
to work; we'd traipse miles, her heels
no hindrance to Stewart Street, a steep walk up

to visit Nana or further on to Ronald Street, up
on the hill where Grandma lived. She'd prepare with care
scones, cakes, and tea, with lemonade for me. Mother's heels
never held her back, always confident, never a doubt
that we'd arrive. Strong personality, strong shoes.
We'd greet everyone on our way. Her friendly face,

recognised by all we met, chatted face to face.
I understood that I was strong, never asked to get up,
to be held like other kids. I knew my shoes
would get me there, my mother's constant care
was never questioned or in doubt.
My memory of her shoes and heels

remain; slender legs, as were her heels.
Unaware of her large tummy, her face
fulfilled with pregnancy, no doubt
that my world was wonderful. Grown-ups
didn't discuss why tummies grew, my care
remained only for her shoes.

I loved those high-heeled shoes and my tan lace-ups.
I'd smile up at her face without a care,
no doubt but trust in this grown-up, impatient to wear her shoes.

Grandma

Warm and cherished on her knee
I'd snuggle into an abundant chest.
Straggly hairs escaped her bun,
clear plastic hairpins slipped out.
Her wrinkled skin looked worn;
aged like a well-loved purse.

Her puckered lips would purse
when I'd get off her knee,
relieved, left swollen and worn;
top heavy in the chest,
unstable on her skinny legs, out-side
propped up by her stick. Her bun'd

slide sideways, seldom neat, though bun
it did remain. She'd take her purse,
stick, milk-can too and hobble out
to the dairy across the road; her knobbled knees
slow but constant. Return to her Chesterfield
couch, where she'd rest her worn-

out knees. We'd read a well worn
storybook while she poked her bun.
In the attic I searched a chest
of hidden treasures. Found the lonely purse
of an aunt who died at twenty-one. On knees
I searched when Grandma was out

of hearing. Found new underwear, pulled it out
of a brand new box. How sad it was never worn.
Grandma won't climb the stairs, her knees
her lone excuse. I returned for a currant bun
and shared a chicory coffee. I showed the purse
that was hidden deep within the chest.

Her tears trickled down her chest,
I hugged her, I've been found out,
and hid the purse.
Returned everything else to the worn-
out chest, including a furry bunny,
whilst she sat to rest her tired old knees.

Grandma festooned her chest with beads and donned her hat, worn
at a jaunty angle, with her hair tucked neatly under, her bun
pristine, we attended church, her purse knocking her buckled knees.

In memory of Penny

When Penny was eight, I was seven. I asked Penny,
'Why don't girls marry girls?' She didn't know.
My mother wisely said, 'Wait and see, my dears,
when you're older you'll decide.'
We played outside most days. Dad had built
a cubbyhouse around the old plum tree.

Some days we'd sit in the flat top tree
and plan our lives ahead. I asked Penny,
'Why do you want to live in a built-up
city, amongst people you don't know?'
'To get away from here. My sisters have decided
to leave when they're sixteen, they want me, their dear

sister, to follow them.' Penny loved her sisters dearly.
I thought they were barking up the wrong tree.
They all loved the movies, that's where she decided
which film stars she would emulate. Magazines inspired Penny.
Her mother made hats with fashionable know-how;
her dreams of movie stars and handsome men were solidly built.

I learnt things at Penny's house. I watched the build-up
before her sisters' dates; plucking eyebrows, something dear
to their hearts. Their father, a disciplinarian, didn't know
about these dates. He was the reason they needed to escape. Trees
covered the Parade back then, where lovers could hide a kiss. Penny
dreamt of romance in the city. In Grade 5 we shared a desk, then Mum decided

I should go to boarding school. Penny's parents went abroad, it was decided
Penny should stay with my parents. She helped me build
a project, 'my perfect home,' in the holidays. I was sad I couldn't go to Penny's
school. She stayed 12 months; loved my parents dearly,
as they loved her. They knew she had potential, the tree
of knowledge lay at university, where they knew

she could fulfil herself, but she resisted, she knew
she wanted to follow in her sister's footsteps and decided
to leave for Melbourne. We met from time to time, my blossom tree'd
existence more privileged. She worked hard and built
a career, lost her first baby due to diabetes. A dearly
felt tragedy. I felt blessed as my three babies lived. I met Penny

again in Ballarat. She knew my parents loved her, that helped her build
the strength to study and to decide to achieve her dearest
dream, to become a teacher. Sadly, my tree rings have outgrown Penny's.

Bonfire Night, 1956

My mother's annual treat
began as sweetness filled the air.
Sugar melted, bubbled fast
cochineal bright red.
No stirring needed, it turned to toffee,
skewered apples dipped and placed

upon an old dimpled tray, placed
high. 'Don't touch!' She did entreat,
'Wait for tonight, you'll get your toffee
apple.' Outside, cold fresh air
stung our face and fingers, I read
alarm across my brother's face. Fast

he told me; our bonfire had burnt down. Fast
and furiously, we began again to place
dead branches. Clive decided to ask Red,
the boy who'd burnt it, to make a treaty,
join us in our undertaking, let forgiveness air,
work towards a great night and share a toffee

apple. We'd make a bigger bonfire, no toffee-
nosed refusal. The whole neighbourhood worked fast
to build a better one. On this May holiday, the air
was cold, the rain held off. Dad placed
some rubbish and old tyres, a treat
we thought, as we hadn't read

about climate change back then. Red-
nosed, we worked throughout the day; no dragging toffee-
feet. We gathered in the evening air
to light our bonfire. Potatoes skidded fast
into coals. Mother waddled from our place
across the road, with an air

of anticipation. Someone threw a cracker in the air
that landed on Mum's tray of crackers; sparking red
rockets flew, Tom-thumbs exploded at her feet. Flowerpots placed
forgotten now, as were the toffee
apples. My mother's waters broke so fast
she left the party to be treated

at Meercroft Cottage Hospital. A deflated air, an early finish, toffee
apples eaten. That Empire Day, red-faced Angus was born fast,
a remnant memory; my special youngest brother, our place and annual treat.

Miss Street, 1962

(after Laurie Halse Anderson)

I hated school. Regulated by a bell.
Woken, Clang, fed, Clang, lessons, Clang,
Phys Ed, Clang, fed, Clang, 'homework', Clang, bed.
Clang, Clackety Clang! Nothing like home, soon one
understood. Enquiring minds suffocated. Bright, crowded
pubescent girls thwarted; a recipe for anarchy. Authority

demanded. Spontaneity smothered by militant authority,
suppressed self-esteem, hope drowned by the bell.
Incessant rules, anaemia, lost identity, overcrowded,
drained spirit, pushed into a murky underclass. Clang!
Bullying, jostling, frustration festered, one
escaped to desperate dreams of suicide in bed.

Caught in locker room with Mother, both crying, bed-
lam on return to school. Miss Street, a thoughtful authority
wrote a sympathetic letter to Mother, describing one
of my tricks; putting salt into a friend's water glass; after the bell
of her departure. My misdemeanour showed that my clang
of rebellion had returned. Miss Street understood. As the crowded

boarding house mistress, she took me aside from the crowd
to her study to suggest some extra tutoring before bed.
The following day I went shopping, not aware of the clang
of disaster that lay in wait. Miss Street, a gentle authority,
recognised my potential. That night the toll of the bell
rang for her. She died in the bath and was taken. This one

sympathetic person gone; corpse carried away; the one
teacher who had belief to help me achieve. The crowded
boarding house gaped, walked in crocodile formation, as the church bell
tolled in mourning. 'Abide with Me' suffused us all. Bedewed
soaked cheeks returned to the world where dispassionate authority
ignored our sorrow. Order restored by the Clang

for tea. No longer Miss Street at the head of table, an empty Clang
resounded. Her trip overseas now a dream. She'd lost her love, her one
true love in the First World War. Teaching with an inspiring authority,
she slept with a dictionary by her bed. She'd shared her hopes, crowded
memories; with her long service leave due; all her plans bedimmed.
Her final moment taken, heart stopping to a silent bell.

Miss Street is remembered with every Clang in so many crowded
places. She saw deeper than most, saw potential in lost ones who cried in bed.
Her authority inspired me to teach in a school without a bell.

Year at Home, 1963

Sleeping in; a year to shed my skin,
an essential act to survive, to discard
the memory of boarding school where
my sense of self was flouted. Time
to read, restore, recover, readjust
and ponder upon life's meaning. Enjoy

the deep rootedness of home. Enjoy
hot toast and marmalade with skin
from our grapefruit tree. Adjust
to touch-typing lessons and discard
anxiety, to share painting time,
life drawing and prepare for where

I wanted to go; Art School, where
my friend Christopher shared his own enjoyment.
He painted my portrait that year, a time
ago. I painted over it, as my skin
reflected my mohair ochre jumper. I had to discard
that vision of myself. Take a year to readjust.

Uncle Henry arranged a trip, adjusted
plans for us to leave in winter, to where
the sun shone every day. He hired a car, we discarded
our winter clothes. My mother came to enjoy
a break away and be a chaperone. My skin
absorbed the sun; a special, joyous time.

We travelled to Rockhampton, a time
to meet pastoralist relatives, inland; just
cattle, that was all they bred. Their weathered skin
had hardened like the herd. Where
we travelled through coastal towns, we enjoyed
the sea and drank mango juice and discarded

thoughts of winter. Coconut palms discarded
large coconuts; fresh new tastes, time
to try new things. I learnt to enjoy
these simple things, not choose the largest dish, just
to prove I could. I paid the cost, vomited where
the pot plants sat, with palish grey-faced skin.

Discarded worries; my time with family well spent, to readjust
to freedom. Timelessness, love encircled, where
I found myself and enjoyed living in my own true skin.

Hobart School of Art, Nipaluna, 1964–66

From a year at home, to find my feet,
I boarded with Mrs Spencer, a kind,
widowed, gentle woman, who mothered me;
she pampered with rabbit pie and fruit cake.
Her lost pilot son's photograph displayed
her sorrow on the mantle for all to see.

I walked up the domain, the sea
lay in the distance. The Gothic feat
of sandstone blocks displayed
massive architectural beauty; a kind
of sanctuary, a wedding cake
art school just for me.

Sky high ceilings awed me.
Life drawing many times a week; see
and draw the rhythm. I shared my cake
at break. It felt like privilege to sit at the feet
of others artists, watch the confident kind.
My gentle touch was on display.

Oil painting on Masonite displayed
our dedication. They taught me
to prime with rabbit skin glue, the kind
that stinks but does the job. I'd soon see
the need to expand to six by four feet,
now too engrossed to eat the cake.

I learnt that turps permeated cake.
Lettering, once a week, displayed
concentrated effort. It was Umberto's feat
to teach the class; he soon gave me
private lessons. We linked up, he'd see
me home and pick me up, a kind

and generous teacher. His kind
mother made a Black Forest cake
for my eighteenth birthday party. We'd see
each other at weekends. He displayed
his Italian ardour, kissing me
in public; I soon found my feet.

The kind memories of art school were displayed.
'I can't eat my cake and keep it.' For me,
my choice: an artist or a mother; both would be an enormous feat.

Peaceful Preparation, 1967

Umberto had recuperated from a nervous breakdown.
I'd stayed in Sydney to console and cajole my beau,
discovered hidden coves in his father's boat, bathed in
warm waters where sharks had been seen, scented
out intimacies beyond my ken, lulled into peace
of mind, reality hit home at Christmas; pregnant

pause, tormented January, was I pregnant?
Torn between two loves, admitting to the other my down
fall. Bert proposed before I knew. Shattered peace.
Engagement dignified my uncertain state. Bert, my beau,
had worked to save for a diamond, set by artisan, sent
from Sydney. His romantic Italian flattery helped in

my decision to be brave. Twenty-first birthday in
the family home, dispelled anxiety, not pregnant.
Party twinned for engagement with friends, scented
candles and pink champagne. Angus, ten, fell down
drunk into the pond. My grandmother said, 'Lovers bow,'
of the necklace given by my former lover, her piece

of advice brought me no peace.
My Gran knew not of complications but, in
her own sweet way, she probably knew. My beau,
she called Albert. Umberto waited in pregnant
silence. She let him know she preferred another, down
right rude, but was she right? I sent

my fiancé cards and letters, he sent
me daily missives. He taught in Hobart. Peace;
I taught up north; weekend sojourns down
south, left me time to sew, paint and plan in
between classes. Molly White, a chain-smoker, with pregnant
dexterity made my dress without her usual bow.

I believed Bert's treatment would prevent him bowing
to his illness, living in a scented
unreality, not aware of the enormity of his pregnant
untold desire to never live in Australia. Another piece
hidden, he didn't want children either. Scholarship in
Italy brought our wedding forward. I felt buffeted in fluffy down.

Indecision bowered me down, not wanting to break the peace.
He came with yellow scented broom, I shuddered, in
pregnant anticipation of wedding hay fever and invidious countdown.

Urbino, Italy, 1967–8

Hoisted high upon a pedestal, she
married young, her eyes so green
blinded by life's complexities, a fairy tale
blunder, the moon no longer made
of cheese. Transported to a far-off land
to be imprisoned as a stranger.

Surviving her splintered pedestal, this stranger
lived to paint, cook, walk and write, she
created stories within her letters, land
of ancient antiquities with vineyards green,
where her devoted prince, imagined, was made
hero to buffer parents with her made-up tales.

Signora Rossini, the landlady, lived her fairy tale.
Her husband in a mental home remained a stranger.
She fattened a pig; slaughtered and made
sausages to survive the winter. She
showed this young one how to glean salad greens;
raw or cooked weeds, picked from the land.

Steamed bitter verdure in olive oil and garlic; land
abounded in weeds galore. Her fairy tale
prince ate rice or spaghetti with these greens;
her tomato *sugo* with rosemary fed this stranger.
With no oven or hot water in the kitchen, she
begged her mum to send recipes of things she made.

Digging snow from the long driveway made
her warm on freezing days. In spring the land
exploded between the vines; red tulips she
gathered in abundant bunches, a fairy tale
gift that lifted the spirits of this stranger
from far away, where memories remained green.

Her husband's frog-like jealous eyes so green,
criticised her wish to please him, she made
him hot chocolate when he appeared, as another stranger
had taught her, in this new and fascinating land.
The more she tried the less she pleased him, the fairy tale
disintegrated. Again and again she

felt his green eyes pierce her, she felt skewered to his land.
Made isolated, she knew it was the end of the fairy tale
she had loved; he'd become a total stranger.

Christmas, Tarvisio, Italy 1967

From Urbino they visited relations,
in Rimini, where His paternal *Zia* Gianna lived,
who had helped them on arrival.
Then to Tarvisio, to see the maternal
Tyrolean *Nonni;* wider families with
male ski instructor cousins. Here

she had a postcard in the mail, so lovely to hear
from Judy. In the same mail, related
news from her mother, Judy had been killed. With
that her tears erupted; her parents lived
so far away, her anchor and maternal
Antipodean support; a plug pulled since arrival.

Anxiety overwhelmed her following their arrival.
Umberto looked so like his *Nonno.* Here,
a scary image of Him in time to come. Maternal
genes had carried through, in looks and related
habits. *Nonno* sat as *Nonna* stood, she lived
to serve him. She couldn't sit at the table with

them for a meal. Instead, she ate in the kitchen, standing. With
encouragement she sat with them until *Nonno*'s arrival.
Nonna bought fat jam doughnut treats. They lived
above the business where tiles were sold. Here
she sat while Umberto skied; she wrote to her relations.
Nonna was a kindly soul but her maternal

heart was wounded. Her only daughter lived abroad, maternal
love was thwarted. Christmas Eve the snow lay thick, with
comfort from Zia Amelia and Zio Ernesto, such kind relations.
A nativity scene set below candlelit tree. Fine feasting on arrival.
They walked to church with cousins; they stayed outside, heard
crowds sing; stood in freezing cold as oldies lived

their traditional lives unhindered. All arrived home to liven
up with grappa and exchange of presents. She thought of maternal
Christmas celebrations in the warm Australian sun, so different here,
though special too. A magic in the snow, heavy with
silence when it lay so thick as it had since their arrival.
She'd been pleased to meet with more of his relations;

though unable to communicate, she yearned to live with
understanding. Her maternal thoughts were dashed since her arrival
here in Italy, she learned he wasn't keen for more familial relations.

Summer Holiday, 1968

Her mother sent money so they could explore.
They left Urbino in their NSU Prinz, a two-
cylinder grey car, a 'lawn-mover' he called
it. They travelled through Austria, to Germany to visit
Ulm, where the Bauhaus had left its mark.
Historic architectural, atmospheric note.

Visited churches, beer halls, all noted
with pride, travelled through France to explore
even more. The ferry crossed to the landmark
of Dover, wandering from one B&B to
another; to London, a cousin visited
then off to Cornwall; on a penfriend they called.

She introduced them to pub crawls, recalled
so many familiar scenes. She noted
the street names of Monopoly they'd visited.
Through Wales up to Scotland, where explore
they must. Her forebears tugged her to
seek out where their deep roots were marked.

Her Scottish forebears had left their mark;
the tartans and marmalade, so familiar. She recalled
the Isle of Seal, Clive had recommended too,
an island attached by a bridge that he'd noted.
Big breakfasts set them up to explore.
In Glasgow, Jenny welcomed their visit.

They ate at a lake far beyond their means. Visited
a friend of her mother's, which marked
a highlight, before they left to go north to explore
Loch Lomond. No Loch Ness monster, so they called
Isa, her mother's penfriend in Edinburgh. They noted
two sisters who repeated each other. Perhaps these two

spinsters stretched Umberto too far, too.
He suffered a meltdown following this visit.
They sat by Henry Moore's dark sculptures, bleak note.
Umberto exploded, a torrid time marked
the last city of Scotland, both ready to call
it a day. They'd overreached their time to explore

and returned home to Urbino, to pack and mark
their last visit, before setting sail for Australia. Umberto was called
back. His resignation noted, before their next exploration.

Baby in Urbino, 1970

Photographs aplenty occupied Him in the euphoria of the birth.
When sated, He became accusatory, suggested she'd been unfaithful.
'She looks Chinese!' This puzzled the new gran who sighed with relief
knowing the baby was His. Her one-night stand in Bangkok on her return
from Australia with a curly-haired Italian steward was her one secret affair.
They'd met again in Rome, where she'd discovered she was not pregnant.

Thank God He never knew of this mis-step. His jealousy was pregnant
with fury, not knowing. He took off to tell His friends of His baby's birth.
Jealousy a family trait, though she didn't know back then; the believed affair;
when his paternal *Nonno* had shot his *Nonna*, believing her to be unfaithful.
Her mother's only Italian, *'Bella Bambina'*, was shared on her walk as she returned
to an empty apartment where the power had failed. To her relief

Marta organised an electrician, who fixed the problem; with relief
Gran poured him alcohol, as this was the custom. Now Gran was pregnant
with questions of where He had gone and when He might return.
After four days, He'd made no contact. To leave hospital, the birth
must be paid for before departure. Gran paid as her parsimonious, unfaithful
son-in-law kept his distance in Switzerland, no doubt having an affair.

Without a translator, she managed; they left hospital to get her affairs
in order. When He returned one week later, it was with relief
He told them they'd return to Australia in three months time. Unfaithful
lover had moved on? She cried when Gran had to leave. Pregnant
with the excitement of going home to Australia to recover from anaemic birth
and introduce baby Francesca to friends and family on their return.

Three months later, she'd packed in preparation for her return
flight. He was to follow. Nappies washed and jammed into travel bags, affairs
in order, waiting to depart for airport. He arrived to destroy the birth
of this idea. 'A cholera epidemic in Turkey!' He shouted, His relief
portrayed. 'I won't let you kill my baby!' Amazed and disappointed, pregnant
anticipation burst. She rang Australia from the city, to tell her unfaithful

news. Between sobs she relayed her broken dream, knowing her unfaithful
husband was in control. She knew now they might never return.
In Italy, a child belongs to the father. With His condition, pregnant
thoughts occupied her mind. His jealousy of baby had altered their affair.
He'd resigned from teaching, maybe a return to Zurich would bring relief?
His abilities were appreciated there; she allowed this thought to give birth.

Selling unpaid white goods for so little, His unfaithful affair,
the broken promise of returning home, she felt no relief.
Fat after pregnancy, depression gave birth.

Witikon, Switzerland, 1970

Beyond the dark forest lies
a foggy battery of muddy graves.
Lichen fills the carved in letters,
words disclose mere birth and death,
while skeletons mask secret lives
concealed beneath the frozen ground.

Sodden leaves blanket ground,
funereal hordes gather, wreaths lie,
mourn those who lost their lives,
freshly buried in their graves,
lament, unite, in untimely death,
place hydrangeas with fresh letters.

The lonely mother rereads her letters
hidden by screened window, watches ground
as baby cries, her sole companion. They await death
trap of His return with meretricious lies.
She strolls with pram by graves
reading headstones, imagining others' lives.

She conjures how her own epitaph of life
should read, short or long, a life of letters?
A sole unattended grave,
for strangers to imagine a life ground
down by deception and lies.
She'd prefer her body cremated, her death

blown and dispersed by the wind. Not death
decomposing, tortured by a tormented life;
etched in stone. Instead, her life of lies
must be burnt to cinders with her letters
creating compost to fertilise the ground.
Not leaving another desolate, lonely grave.

Her ashes will nurture life beyond the grave,
fertilise plants, regeneration follows death.
Revitalised, replenished ground,
vibrant flowers to enrich future lives.
Lost mixed messages from her letters
decompose with their lies.

Cremated in death, her story lives
beyond her death and burnt letters,
grounded in sestinas, in a book they lie.

Cabbio, Switzerland, 1970–71

She listened to the frozen silence,
baby smiled and nuzzled closer
within a clockless infinity.
Fed, slept, walked and wrote,
no other form to communicate.
Waited for letters that didn't come.

No phones to tell her He won't come.
Isolated, a village where silence
persisted, where she couldn't communicate.
Night noises reached close
from smugglers' track below. She wrote.
Snow encompassed infinity.

Dreams dredged deep infinity,
uncertain if He would come.
He doesn't write.
She boiled the kettle to break the silence,
cosseted her sleeping baby closer,
waited for her to communicate.

Gurgles tinkle and they communicate,
she no longer feared infinity.
Bursts of sunshine brought warmth closer,
Smiled at her darling child, 'Here I come.'
They'd walk together, absorb the silence,
life would wait for her to write.

Church bells bellowed. Write
tomorrow, then maybe He would communicate.
They'd walk and pierce the silence,
share the scant blue of infinity.
Silent sled; her baby overcome
as shy deer drew closer.

Circling mountain, home grows closer,
ideas spring forth of things to write;
hope for days of laughter, still might come.
Wait, wait to communicate,
she'd share her difference of infinity,
absorb His absence in the silence.

Closer, baby's joy communicates.
She'd write about infinity
of difference. He'd come and break the silence.

Effretikon, Switzerland, 1971–72

Married to an illegal immigrant worker, they'd had to flee
from a census in Witikon. Lucky to be saved by Aldo, a friend
who gave them sanctuary in his Cabbio cottage, where
they'd hidden in the mountains; there she found respite and retreat.
Aldo's antique treasures hidden in his fourteenth-century villa, four storeys high,
opened her eyes to a wealthy world; she was forever thankful for his help.

Devastated to leave Cabbio, though this permitted medical help,
sadness swept another solitude. He chose to flee
to Effretikon, a suburb of Zurich, living high
up amongst Swiss German speaking strangers; her baby her only friend.
Again, she made a home, propagating African violets in this new retreat.
She found others who spoke English, friendships were born where

she'd share with Friends at Quaker meetings. That's where
she met Philippa, married to Manfred, and their baby Rupert. Help
and hope restored, they met in each other's apartments, a treat
to share weekly lunches, the children content to play; no need to flee.
They discovered Zurich together, a friend
with whom she shared artistic endeavours that left her on a high.

Annamarie, a Quaker weaver, invited her to aim up high.
Apprenticed four mornings a week to weave, this was where
she met other artists and Annamarie became a close friend.
Annamarie took up her psychiatric practice half days, and with her help,
she faced her difficulties at home without having to flee.
Instead built new skills to regain her own secret retreat.

One weekend, when He'd gone to Italy for his own retreat,
she and Fran left for a commune to escape his return, but the high
of group sex wasn't on her agenda; again, she'd had to flee.
A solicitor advised her of her rights: return home, stay put, and where,
on his return, He'd have to give her half his salary or be deported. This help
gave respite. He moved out to stay with a friend.

Annamarie continued to interpret her dreams at friends'
rates, that helped her remain sane. He lasted two weeks in retreat
and returned, pleased with her. No longer helpless,
she had made changes. He praised her efforts to learn High
German now that she could afford lessons. Never satisfied, where
or when, He again sought another move from the irritating fleas

of friendships in Effretikon; removed them to on high,
away from outside interference. Retreat to Scheuren, where
help came from Friends, who had been Jews, who too, had had to flee.

Scheuren, Switzerland, 1973

Close to nature in her youth, she felt a tear,
a hiatus between the energetic sea
and these leaden lakes with depleted life.
A need for fire, she lit her candles,
cooked fondue to warm them through.
Grew amaryllis, her token garden.

A rose from Quaker Lisa's garden
undid her and brought a tear.
Never to grow roses through
the seasons; nostalgia for the salty sea.
In pure air; no smoky fires, just her candles
to deflect the sterility of high-rise life.

He chose their furniture, white settees, life-
less, not considering child's grubby garden
pot plant hands. Francesca loved the candles,
learnt Swiss German with occasional tears.
He, in Zurich, designed typeface, never to see
or notice how His moods threw

her. Swiss rules were strict: vacuum through
nine to twelve or three to five; a regimented life.
Residents allowed one laundry day a week, to see
and share the washing machine below garden
level, dark drying rooms, always tearing
hurry to complete laundry in a day, home by candle-

light. She made muslin curtains, cheap as candles,
unsuitable to keep out zero temperatures through
winter, but they created delicate light as she lay in tears
after a miscarriage, ignored by Him. She mourned life,
Francesca's lost playmate; together in the garden
in her dreams where they might be by the sea.

Brindisi brought comfort, holidaying by the sea.
Yes, she romanticised a perfect family and held a candle
for another baby, believing this would bring garden
of satisfaction. Once again, *incinta*, through
better times, she believed in this new life.
She gave Annamarie her laden cumquat tree; leaving with no tears;

left a symbol of fulfilment. New life lay across the sea, where He'd follow through
with His plan to be a student. She'd light candles to reflect the joy of a new life,
where they'd speak English, have a garden and await Rebecca's birth without a tear.

Ufton Nervet, near Reading, England, 1974

Dried fruit and nuts to survive the train
journey, evacuated on rough ferry ride.
She slept with her child in smoke-pervaded bedroom,
strange opaque memories, so long past.
Arrived at Wises Firs bedraggled and tired,
wheelless suitcase opened; His wine a hidden weight.

Discovered the rented house and garden where weighted
case off loaded; her nesting urges all in train.
With scant to lose, exhausted sleep, their tired
bodies, hers six months pregnant on the downward ride.
She was woken by penumbra ghost, fright passed.
It wordlessly said, 'All will be well.' She lay sleepless in her bedroom.

The spiritual vision pervaded her bedroom,
gave strength to her penurious weight-
filled heart, grateful she'd survived the past,
met new and interesting people. She trained
her thoughts, prepared for the next exciting ride,
uncertain of His challenging return; over-tired.

Three months later, He came. Her parents arrived tired
from Australia, observing He no longer shared her bedroom.
Baby coming, He refused to rise. Midwife took her to ride
in her plastic seated Mini. Her mother, weighted
with worry, stayed up to hear of baby's birth. Born train-
fast, she felt undaunted by her troubled past.

Home the same day, she cherished Mother's past
and copious skills. She came downstairs tired,
with disbelief. The carpet gone! He'd caught the train
and changed it for another. They fought upstairs in her bedroom.
One day absent and He'd thrown his weight,
believing He could control her in another idiotic ride.

Angry. No longer would she let Him ride
with hobnailed taunts to crush her; her past
passivity gone, time to heft her weight.
Gone, her days of being too tired
to fight. He looked elsewhere for bedroom
fun and left her pondering on His train

wrecked rides of past affairs, leaving her feeling tired.
She found a letter from Brigit, a pregnant past girlfriend, in their room.
He denied it but reality weighed heavily. Major changes must be put in train.

Ufton Nervet, 1974–75

The birth of Rebecca made time fly.
Home-made Christmas presents, her usual feat;
she cooked, knitted and sewed. Made
navy and red woollen garb for baby Beck and a gift
for Fran, a yellow stick hobby horse with
navy blue head and mane of flaming red.

She watched Fran proudly ride, her cold cheeks red
as she pranced with shrieks up the street, fly-
ing snow billowed as she sought out her friends with
a wave to them inside, they'd point at her frozen feet;
from behind glass they'd show their extravagant gifts
of things that were bought, not home-made.

She started portrait painting classes and made
some friends. The class planned a red-
letter day to visit London and view the gift
of a Turner retrospective. She knew the day would fly.
She left Fran with a friend, baby with Umberto, a feat
not done before. She left with elation and trepidation, with

heart hammering, she boarded the train with
excitement for a day of freedom. She'd made
her mind up to take the chance, a feat
not taken lightly. She soaked up the Tate, read
and viewed with respect, let Turner's colourful life fly
by with his exquisite landscapes, a virtual gift.

Arriving home, her neighbour distraught. Her gift
of trust no longer there. Their house was locked with
distraught baby left alone inside. She frantically flew
to Becky's cot to grab her in a hug. He'd made
her furious, how could He leave her, screaming red?
Trust gone. His dishonest words veiled His worthless feat.

Plans anew, she knew she couldn't get cold feet.
He, a student, lived the life of a gifted
boy not wanting responsibilities. She read,
immersed in other worlds, while children slept. With
no TV to distract them, her children made
life bearable. She'd keep them safe; they'd learn to fly.

Her feat now to refrain from spurring Him on with
antagonistic words; her gift to herself. She made
sure she'd no longer show red anger, or let Him squash her like a fly.

Ufton Nervet, 1975

Umberto returned from abroad where
a secret love had restored His deflated ego.
She, once again, brought Him back to reality.
He refused to eat her prepared meals, instead
controlled her with diets galore, with purchases beyond their
means; embarrassed guests when He'd not share his wine.

His studies resumed. He relayed with a whine,
the need for a break, He'd stay with a friend where
He'd decide in six months if He'd like to return to their
home. She gathered her shattered ego
and said she'd return to Australia instead,
where her family lived in a peaceful, safe reality.

He tried to seduce her, his reality
duplicitous, she refused his wine.
He moved out without further ado. Instead,
He'd be free of responsibility where
the freedom of student life fed His ego
and he could proudly show their

daughter, Fran, to His uni friends in their
quarters, out of touch with reality.
Forgetting to pick her up when His ego
got in the way. She tried not to whine,
but watched when He didn't turn up where
preschool children were collected. Instead

she'd take Fran home, knowing His fickleness, instead,
as the school had rung her, just once. Their
routine restored when He went overseas where
His Swiss love awaited. He lived in an unreal reality
where He did what he pleased; love, wine
and song, being treated as a hero, his ego

engorged. Her worth reduced, her ego
battered, she worked on her project instead;
to return to her homeland. No need to whine,
her parents had foreseen; by buying her a car, they'd secured their
grandchildren's freedom; left a reality
for her to sell, to pay for their fares. Thank God for where

ever His ego was replenished, far away from them.
Instead, they'd rebuild and evolve by living fully, find their own reality,
where they'd be surrounded by loving, reliable friends and perhaps some wine?

Secrets of a First Marriage

Attempted murder, suicide, His grandparent's story,
discovered only after separation, when my in-laws came to stay.
So many hidden family secrets.
Lies by omission, living under sombre shadows
of painful past. She was unaware of unseen
bruised indentations on His broken psyche.

Like damaged deer, this soft, tender psyche,
shocked by treatment, another story
to resurface at a later date. Damage unseen,
hurt inflicted, children's plastic toys removed, nor stay
her dresses of certain colours, gone into the shadows
never to be found again. These secrets

never were discussed. He, in command of funds and secrets;
she became His mere possession. Doctors tried to heal His psyche
to no avail. They moved whenever she made a friend; just His shadow;
sixteen homes in eight years. Left at home to dwell upon their story.
His yen to move a constant, even in Italy they couldn't stay.
No one understood his need for untrodden scenes.

Seasonal summer highs, immersed in other scenes,
International conferences, He'd speak. The secret
of His lows she coped with. He'd stay
in bed and hibernate. Swiss psychiatrists helped His psyche,
allowed Him to mull his story
but never fully remove the shadows.

Depression saturated shadow,
a murder of crows against the snow, unseen.
One thread unravelled stories,
homeopathic droplet of truth revealed another secret,
heaviness imposed upon her psyche.
Taxidermist destroyed her self-esteem. She couldn't stay.

Pregnant, she left for England with their child to stay
near Reading, out of his shadow.
Her parents arrived before him, His unstable psyche
revealed. He slept in lounge and refused her food, the scene
of troubled married secrets
exposed. No longer could she hide their story.

Parents stayed; birth of a second beautiful baby girl completed the scene.
When parents left, He refreshed with fresh flesh; leaving shadow-steeped secrets.
She returned to Australia following last psychotic episode that ended this story.

Between Marriages 1 & 2, 1975–80

Loneliness struck with a forceful swipe,
tears would emerge when I'd least expect;
the newsagents, fruit palace or in the bank,
shame and embarrassment covered my face.
Alone at night when the children slept
I'd heartily sob with remorse and regret.

Marriage bonds broken, too late for regret.
He'd found another, the past I'd swipe
clean, sew patchwork as children slept.
To find another is what you'd expect.
Clothing and construction, a big step to face,
made dresses, that helped save me at the bank.

Marimekko fabric came to the fore, bank-
holiday glamour, silence restored; far from regret,
I celebrated Little Gallery's decade; faced
opportunity, drank red wine with a swipe,
met Peter, recent widower; we hardly expected
to click, yet we did, together we slept

after joyous party. Miss Stoner, my teacher, slept
in my back bedroom; students galore, banked
Christopher's exhibition success. Least expected,
Edith Piaf's voice resounded, *'Non, je ne regrette
rien'* that filled the house, foreboding a swipe
at a future we couldn't foresee, or were willing to face.

Four years of contentment, I refused to face
fool's fantasy, together we slept
in harmony. Grandparents milled, then with a swipe,
Umberto returned with his next wife. Bank
of longing, the children's dad had regret.
Girls hardly knew him, what did he expect?

Fran five, Becky one when we left. Expect-
ant grandparents rejoiced when we met face to face.
Australia was home, we felt no regret.
Umberto was jealous when he learnt that I slept
with another. He refused maintenance but his wife, Sandy Banks
made him pay. In those days there were no cards to swipe.

I knew I'd have to move on, to study, whilst Peter slept.
Leave, to further my education and not be dependent and bank
on his kindness. I regretted hurting him with that final swipe.

Hobart, Nipaluna, 1980–81

Desperation, insanity, call it what you will.
I'd left the comfort of the known,
two small girls in tow,
for university; a better life ahead.
Quaker yearly meeting presented
a lone mysterious spirit.

This time it was my turn to be spirited.
Love comes in many guises; weak of will,
smitten. Charming, tall, he presented
perfection to my prayer. The unknown
intrigue drew me; instincts numb for what lay ahead.
Three days, my judgement lapsed, I became in-towed.

He departed on two weeks holiday, I packed and towed
our needs to Hobart, where I cleaned with spirit
that filthy flat, left desecrated by dogs and rats. Ahead
of his arrival, windows sparkled, curtains sewn, my will
to please still in inception, a new life, yet unknown;
became distracted by gifts that were presented.

Parents' acuity clear, what this new encumbrance presented;
unemployment, shoes tucked under bed, a drowning undertow.
They questioned, 'How could you not have known?'
Too late, the parsley seed had germinated, spirited
and growing; decisions; wedding plans. My fragile will,
an incessant see-saw, plunged into plans ahead.

Children settled, yet more unwanted moves lay ahead.
Thrilled to be surrounded by siblings and cousins. Presented
with a healthy baby boy, life resumed its crazy pace. Will
postnatal depression suck me into grasping undertow?
Survival on a student pension took strong spirit,
immense courage, to cling to the vision of a better life. Unknown

to me he thwarted my opportunity to teach at TAFE. Unbeknown,
he'd thrown away my plan; accepted a Cairns teaching post ahead;
his ruined chances to teach in Tasmania revealed. Spirited
away again from loved ones. He sold my diamond ring for presented
transfer bills. My parents and children cried, a taut tow,
we followed him against our better judgement and our will.

Cairns, an unknown paradise, presented
new adventures. Life ahead, with sand between our toes,
refreshed our spirits. We began again with renewed will.

Cairns, Queensland, Gimuy country, 1982

Hobart to Cairns through fluffy clouds,
we stepped into blasting heat. Encircled
by vibrant hibiscus, luscious verdant plants.
Sonorous birdsong looped through perfumed air.
Scared of cane toads, we waited
for chattels to arrive at our motel abode.

Motel pool enticed us out of our abode,
I felt as light as feathered clouds
whilst swimming; sleep could wait.
Children attended the city school encircled
by new students; adjusted to warm air
with no respite in the night. I planted

seeds of hope, we'd dig and plant
a garden. We moved into our new abode.
Ian had PD workshop and travelled by air.
I was ill with Dengue fever clouding
my ability; feathered Red Indians encircled
me in the passage as I awaited

his return. I applied for a job, and waited.
Interview granted, we both attended. A chance to replant
our family with up to eight in care, encircled
my dream goal. This change meant a larger abode.
Seven interviewers at the Group Home, a cloudy
prism troubled Ian. I felt composed, an air

of knowing. They questioned, I joked. The air
relaxed; they checked to see if I was kidding. Weight
lifted when they could see the joke. Any clouded
doubts disappeared with our success. Transplanted
yet again, this time to a two-story red brick abode,
a garden that large trees encircled.

I observed our three children encircled
by displaced children; I knew it could go either way. Air
of trust, the record Grease's lilting rhythm satiated our abode,
accompanied by the thrumming trampoline's incessant thump. Wait;
another change of schools, another crisis. Plant
new friendships to douse the clouded

sorrow of lost new friends. Encircled with new friends we waited,
trepidation in the air. Rapport grew strongly, like thriving plants.
Frangipani perfumed our abode, birdsong pervaded skies, now cloudless.

House mother in Family Group Home, Freshwater, 1982–3

First girl cowered, covering head,
when confronted with the stolen purse. A
hug to show it wasn't her, just
her behaviour was in question.
One boy, a habitual smoker; how to respond?
Outdoors only, not in bed.

Quick decisions on the run. Bed
time and boy heard crying overhead.
Not missing home tears, as first thought. I responded
quickly, appendicitis; off to hospital, a
successful outcome. Kids in crisis, no question:
emergency overnighters and long-stayers. Just

a band-aid solution; justification
to help avoid conflict or incest; all required a safe bed.
Meeting others helped children question
better options for a safer path ahead.
Girls found friendship, setting goals for a
fuller life. Francesca led, the girls responded

and followed suit, doing homework. All responded;
painted, sewed, cooked and swam. Just
when things calmed down; small crisis, an
invited sibling arrived with nits. Beds
stripped and laundered after all heads
treated; outside grooming, outing questioned.

Who'll be the 8th to come? The question
remedied with a pre-adoptive baby. All responded
with compassion. Eight in place, teenage girls' heads
filled with maternal instincts. They fed, burped, or just
shared stories; made a book for baby's bed-time;
our shining jewel, loved to pieces, a

deliverer of trust and hope. A
gift to lighten the load of broken lives. Questions
of stealing, smoking, incest, violence embedded
in their tragic lives. New found hope in response
to the innocence of baby's smile. Plus, two-year-old Simon just
added natural love of fun and laughter, helped them to look ahead.

A blessing passed between us, girls responded,
aware now of the need to question injustice,
ensured safety in bed; dreams of love and hope suffused their heads.

From Freshwater to Clermont, 1984–86, Yiraganydji territory to Wangan People's Land

On the hillside we bought a block of land,
planned a cyclone-resistant home,
our first. Rented as we watched building course
ahead. I worked in a bookshop. Children settled in.
I planted lychee, hibiscus and pawpaw trees,
made friends with neighbours.

Christmas party with closest neighbours,
drank cumquat brandy on fertile land.
Magnificent mango trees,
fruit for hummingbirds and bats; their home;
nature invigorated by the clime. I chose curtains in
yellow and white diagonal stripes, that coursed

through living areas. Lifeline course
completed my plan. Palm fronds, friendly neighbours,
warm walks on beaches, picnics in
the bird park where cassowaries roamed the land.
Settled in, close to the group home
so girls could keep in contact, chat under trees.

Eighteen months later, enjoying the trees,
our chosen life shattered. Ian changed our course.
He'd accepted promotion, which meant changing homes.
A move to a mining town; Clermont, new neighbours;
Daintree Street without a tree. Prickly bindies covered the land.
I objected when I saw a gun-toting neighbour fight in

the street. We moved nearer the school in
a more settled area where there were young trees.
Starting again, I planted sunflowers and tended the land.
Francesca left for Sydney to complete her school course.
Her best friend, Erica, our Cairns neighbour
was killed. A school bus accident on her way home.

Fran would have been killed if we'd stayed at that home,
so, the right choice had been made, in
retrospect. Becky quickly made friends with the neighbours
and invited them home. Simon helped me plant trees.
I turned forty with no friends around. Disappointment coursed
like the red-backs in the sandpit that infested the land.

Now a waitress in Clermont; our new home, I missed rain and tall trees.
A friend and I painted in a group and joined a public speaking course.
We left those neighbours the following Christmas for Canberra, Ngambri land.

Canberra, Ngambri land, 1987–90

We visited Angus and Trish when they exchanged rings,
Francesca stayed in Sydney to complete year 12. We
lived, again, in a motel, to adjust and find an address.
I enrolled at university to complete my degree. Found the right
house; waited for tenants to vacate. Sold our first home – that tore
my heart. No going back. New schools in walking distance,

once we moved in. Ian worked at Koomarri, car distance
away. Simon began primary school, his favourite thing, play. I'd ring
my parents in Tasmania. My father had cancer, tears
didn't help. They visited. My dad loved the Science centre, we
dropped him off there when we went to the shops. 'It's right
that you're living so many metres above,' he said. 'This address

will be safe when sea levels rise.' He liked to address
ecological problems; a step ahead of his time. With distance
shortened by air travel, living closer felt right.
Ian settled into special school. I loved my studies, ring-
binders abounded with a trusty old typewriter. We
adjusted to climate and surrounds. I was in a tearing

hurry to study and achieve. Now time to tear
myself away from the television and address
my study needs. Beck studied German, her *Nonna*'s tongue, we
learnt together. Speaking German reduced distance
to *Nonna*. We bought a cat who took the ring
out of her bell. We locked her inside at night for the rights

of the wild life. My degree completed. A right
turn into relief teaching, then art. The third placement tore
me from high school to a special school. No more ringing
in the morning for pot luck placements, a six-month term addressed
my needs. Working with Ian, we travelled the distance.
He had other ideas at home; there were problems, we

disagreed about discipline. Beck incessantly sent to her room. We
decided to separate when he accepted a transfer, this felt right.
Permanency for me at the Woden School, driving a distance
further from home. Ian left for Jervis Bay, Yager country. Tears
were spent as he insisted Simon go with him. I had to address
this option but the final decision was made by a ring.

They moved. Two weeks later, Simon wanted to come home, rang in tears.
We agreed he had the right to address
where he wanted to live. His dad travelled the distance and could always ring.

Canberra, Ngambri land, Devonport, Tiagarra, 1991

Teaching at the Woden School
was like unravelling a
ball of string. Unexpected insights
learnt from students every day.
They showed acceptance and perception;
remarkable staminas to sustain.

My father couldn't sustain
his strength, his time for schooling
had dwindled. Forewarning perceived,
I flew business class, the last seat; as an
unexpected arrival. Friday
night and he'd just arrived, a sorry sight

from hospital. Surprised by the sight,
'This dying's an expensive business!', I sustained
a laugh at his forthrightness, a day
to be remembered. Two days off school
left me four days to immerse myself in a
time capsule, hear Dad's perception

of the past. We read his mother's perceptive
diary, no mention of her illness but an insightful
look at a cultured farmer's wife, a
life spent looking after others; sustaining
her gift of literacy. She'd been a school
teacher in her youth, reading every day

and nurtured by her love of books. Days
spent writing poetry, her perception
constantly aware, helping her own school
children to not lose sight
of a wider world and sustain
their love of nature. Organic gardening a

natural thing back then; a
lesson learnt, one that Dad continued every day
to practise, and sustain.
On climate change, he spoke of his perceptions,
of tides rising and the need to keep sight
of how we can help further school

others in living a fuller, fruitful life. I loved his perceptive
knowing; he sold solar heating in the seventies, his daily insights
spoke of how to sustain our planet and nurture young ones in our schools.

Easter Passing: Little Italy,
Tchukarmboli, 1998

Bleakness as the weather changed
to sadness; Umberto had hanged
himself; not to be found for a week.
Black blowflies had filled his window
alerting his neighbours of something wrong.
Roberto, his younger brother, was the first

to arrange a cremation, the first
before three memorial services. I changed
to be with his widowed mother, so wronged.
We gathered at her apartment and hung
together, looking at the church from her window,
before the service at the end of the week.

A Catholic service late morning, weak
Sydney sun struggling to shine. Sandwiches and the first
flower arrangements veiled in the gloom of window's
shade. Umberto gone and everything had changed.
Sandy was there, Francesca too, we hung
about, but the pleasantries felt wrong.

How had it happened that he'd felt so wronged;
to end life alone, all hope gone, Easter week,
too much to bear. An image of Jesus as Umberto hung,
gone. Misdoings now complete, my first
relieved thought. Ideas he'd discussed, now definitely changed.
He imagined two ex-wives in his window

of unreality; wanted us to live at Little Italy together. A window
definitely closed. A Quaker service was held and, right or wrong,
we travelled to Lismore to face what had to be changed.
His rented abode crammed full of stuff for his new venture; weak
in its concept, never to be fulfilled. The carpet rotted outside; first
step inside, a pungent smell permeated every object where he'd hung.

The paint work half done, he'd hung.
The stench bore into every pore, the window
of his soul revealed. We scattered his ashes where he had first
stayed at a Buddhist farm. We met his latest girlfriend, wronged;
she claimed his car that symbolised his weakness,
it couldn't reverse, nothing had changed.

Umberto remains forever young, he hung, was it wrong?
Depression darkened his window, insight weakened
hope for a better future that he'd, at first, thought he could change.

Secret Reality

Sacred Canberra night passed.
Placed mirror with candles either side,
sat crossed legged in total silence,
gazed at mirror as candles flickered.
Reflected face returned a stare.
'I am that I am', a mantra thrice repeated.

Following night this was repeated.
Hours of meditation passed.
Gazed at unemotional stare.
Frozen, wrapped in blanket, side
ways look, the candles flickered
enveloped in a deadly stare. Silent

children well bedded. In the silence.
no distractions, determined to repeat.
Third cold night, air created candle flicker,
mantra repeated thrice; past
believing this would work, glance aside,
image disappears, no lonely stare.

Miraculous, blank mirror stares,
numb in silence.
Inexplicable, blindsided.
Occult mystery, repeated,
I'm not there. The strangeness of the night
unfolds as candles flicker.

Awareness of ephemerality flickers.
Euphoria; visage no longer stares.
Transience imbues magic in the night,
magnified in dark chilled silence.
Who'd believe this if repeated,
I hardly do myself. Beside

solicitous questions, inside
intimate illusion, flickered
thoughts reverberate, repeat.
I know the mirror's enigmatic stare
exhilarates empty silence,
reflects within the shadows the wonders of the night.

A secret gift bestowed, beside the stare
that vanished, flickered a new reality into the silence;
relinquished body, soul's future repeats from many pasts.

Birthing Biography

Birth and share a sestina through
sharp painful contracted thoughts
as a bird pecks to fracture shell.
Cut cord of complacence, allow
released sticky sap to flow,
wallow in bloodletting free fall.

Swaddled words of childhood set free,
unspoken love received through
actions. Care in abundance flowed.
Privileged first decade, thoughtless
trifles accepted, free play allowed.
Sought, found and gathered seashells,

but for those collected shells
to disappear. Under bed free
of detritus, accepted, allowed
other small collections. Through
beach combing, gathered thoughts
multiplied where sea creatures flowed.

Boarding school inhibits flow,
restricted, repressed, hollow shell
of a child lived with solemn thoughts,
hankering to summon freedom,
to return home, to tread water through
hormone upheaval that allowed

steps to plan for art school. Allowed
three years of fulfilment to flow.
Awakened senses. Experience through
art left fragility, broken-shelled
expectations; should one be free?
Deep desire for a family, these thoughts

incessantly tore at thoughts
instinctual, maternal; allowed
marriage to art teacher. Free
spirited, before foreign language flow
disrupts dreams, left to hide inside shell
concealed. Discovered too late through

deception; loving thoughts had stopped flowing.
Allowed home with offspring. Protective shell
cushioned comfort, free to unburden injustice lived through.

Changes Afoot, 1999–2000

Christopher invited me down
to the opening of the gallery's
new addition. He took me back
to his cottage in the grounds
where he was caretaker.
He cooked boeuf bourguignon as I made fire.

Flushed with warmth from fire,
he asked which bed did I want to doss down
on. My choice was made, he'd take care
of me, after he'd asked another question. The gallery
was beautiful and we both loved the grounds.
We shared Christmas in Canberra, back

to discuss old times. The children backed
our future plan. We shared New Year fire
works that symbolised excitement, ground
covered for our plan to marry in June. Down
to Tasmania we shared time. Gallery
faxes came each day with jokes from my caretaker.

My house was sold with time carefully taken
for my long service leave, to fit back-to-back
with our wedding plan before GST at the gallery.
Jokes and letters enhanced our lives, both fired
up in readiness for change. Winding down
at school meant sharing my art room, grounding

new staff; me sharing other areas as ground
work went ahead to replace me. With care taken,
time would fly, to pack up house and drive down,
with Simon to share the driving. He would move back
but worked in Melbourne to begin. Fired-
up we travelled down, me to McClelland Gallery.

Chris's parents were there at the gallery,
we spent time looking around the grounds,
and sharing meals before the fire.
Chris took me to meet the staff, as caretaker
he kept an eye on twenty acres, back
then, before it doubled in size. Down

at the gallery on the Mornington Peninsula, being caretakers
exhilarated. Looking after grounds with freedom at our back,
we lit fires to clear the blackberries and finally settled down.

Wedding Day, McClelland Gallery, 30.6.20

Margarett, Chris's Mum helped me choose
the flowers to make small arrangements
for each table. A knock on the door
surprised; I thought Fran had changed.
Instead, it was Rebecca, who we thought
was overseas. Our homeling

joined us, so all of our five children were home
to celebrate our wedding. A better gift I couldn't choose.
My father was the only one missing, I thought;
plus my brothers, we had kept arrangements
small as Clive's wife, Leonie, wanted us to change
our minds, determined this door

shouldn't open. Instead, we threw this door
wide, as we felt so at home
together. Why should she try to change
and thwart, impose or choose?
It was our concern, an arrangement
to suit us, not her, we thought.

Sad that we couldn't share thoughts
of joy, Leonie had shut that door,
we continued with our arrangements.
Dame Elisabeth's magnanimity let us choose
a three-bedroomed open-planned home.
Exciting times, a big change

from the cottage. Blessed in blue, changes
lay ahead. We walked over to the gallery, thoughtful
staff attended. The director's brother, Andy, chose
and performed the service in the French Gallery without a door,
a simple service shared with friends and family, our home
was blessed with these arrangements.

Sirens blazing, yellow suited firemen arrived, an unarranged
action, due to a simple kitchen faux pas. It changed
the quiet atmosphere to one of celebration. Champagne and homely
food was passed, not giving another thought
to those we missed; enjoyment filled the air. Out of doors
the moon shone brightly as Wise Choice's

delicious meal arrangement satisfied us all with pleasant thoughts
and elegant sufficiency. We danced as Chrissie played piano. My life changed, doors
opened wide, home with Christopher, the best friend I could have ever chosen.

Mother, Devonport, Tiagarra, 2001

Leonie rang to tell me it was time
to return to Devonport. Mum
was in a bad way. She refused
the ambulance to hospital and I said I
would take her. Strips were torn
off me on arrival, a wrong decision had been made.

My brother, Clive, had made
arrangements for Mum; to lose no time
to move to Hobart and sell her house. She was torn,
her heart remained in Devonport. Mum
had been born there, her friends were there. I
saw she didn't have the strength to refuse.

Her house was put on the market, she couldn't refuse
the offer. A new unit in Hobart bought, decision made;
fresh paint and new carpets were going in. I
felt sad for her as she didn't know how much time
was left, or whether she was well enough. Mum
had been such a strong character, now she was torn,

dispirited. She lay in the Mersey Hospital, memories tore
into fragments. She'd done her nursing training there. She refused
to give up so soon. Family visited on her last weekend. Mum
alert, bright red cheeks, breathless, smiled and made
small talk, pretending all was well. Three weeks and her time
had drawn to an end. Fran left for the airport and, when I

returned, Mum was in a coma. Such strength. I
was glad she wouldn't make the move, her tattered, torn
body had petered out. I thought of things she'd said, 'Time
to sort things out. Don't throw those margarine containers.' I refused,
placating; hoping she'd understand the decisions made
when the skip arrived. Ice cream containers thrown too. Mum

had collected every one. Clive came up to sit with me, Mum's
last day and night. Not being in the right frame of mind, I
sat, as protocols flew out the window. We tried to make
arrangements. The funeral notice, an incorrect form. Torn.
Instead of making the wreath I let the undertaker arrange it; I refused
my responsibilities and regretted it in time.

Mum's funeral was huge, my heart wept, tears tore,
I saw spears of pink gladioli; mum's nemesis; her wedding flowers became refuse.
Solace; she'd told Chris, 'She's your problem now,' that made us smile at the time.

Traditional Wife, Pearcedale, Bunurong Country

Gardening, no end in sight; leaves
to rake, spent flowers to dead-head, check
aphids, weed, hoe, prune and mulch, raise
spring seedlings, continuous compost care,
establish a herbaceous border, plant trees to bring
the bees, contrive to propagate and synthesise

methods to save time and labour. I synthesise
house work, a more complex task as others leave
a trail of debris with incessant dust that brings
a need for patience and a mind elsewhere, more checks
for lost socks under beds, hospital corner care,
wash and iron and put away. Tidiness can raise

the spirits, a tight regime if I'd raised
my offspring to do the same. They synthesise
their work and home-lives, show capable care.
A joy to see their evolution to adult-hood, then leave;
the task then remains to educate my partner, check
that he is in the loop, my job restarts to bring

him up to speed. Entice and cosset, bring
breakfast in bed so as to escape the house and raise
a song, to gather apricots and check
for blackbird pecked fruit to stew, synthesise
a compote of mixed berries that leaves
a memory of attentive loving care.

To compare gardening with home care:
weeding equates to tidying up; putting things away, bring
weeds to the compost bin and layer dead leaves
with the straggly greens, sprinkle layers to raise
the temperature with blood and bone, then synthesise
dolomite every second layer, check

the moisture content, let decompose, check
again in a few months' time. Inside the home, as carer
I'll iron and fold, place pressed garments, synthesise
colours with an artistic eye, add a pouch of lavender to bring
a fresh surprise when digging deep, to add aroma, to raise
a memory of summer on a winter's day, that leaves

one to check one has a fragrant hanky before leaving home to bring
a smile. A trad. wife is a caring home-maker, who in turn raises
the next generation to synthesise and compost all their leaves.

For Christopher, 30.6.19

Sleepless in Pearcedale on our nineteenth anniversary, I make a list
of things for which to be grateful. Chris, my creative husband, kind,
helpful on a good day, a raconteur, a mimic, holds fascination with the news
and relays the interesting bits to make me laugh.
We discuss our sister-in-law, who predicted our marriage wouldn't last
six months, at brunch at Le Gourmand Café, our favourite place

and how we'll celebrate in our nineties, return to our favourite place
hoping we'll be healthy and able and not begin to list,
knowing that we have a lifetime of memories to last
us into the next realm, where we imagine meeting those who've left, kind
of crazy notions of our idiosyncratic forebears who always make us laugh.
What unbelievable people; some even made the news.

Our children don't remember our special day, their news
is more important; we remember they were all single back at that place
where we married, at McClelland Gallery. So much to laugh
about then, when we lived in the caretaker's cottage with a list
of things to achieve and where we found a kind
of peace in planting trees, but good directors don't last

forever and we made the most of McClelland until the last.
Then we started again in Pearcedale where the news
of our transfer into a life of retirement and being kind
to ourselves, began. Having to plant a garden at our own place
and pace took precedence; where Chris can paint. We again make a list
of the things that bring happiness and make us laugh.

Now we have nine grandchildren to make us laugh.
Two to stay in the holidays, a joy which brings memories to last
and provide conversational discourse with ideas to list
educational changes with generational verve and updating news.
We listen with grateful interest in our own remote place
and consider how it was for us with housebound mothers, kind,

there for us. Life was free in those young years. Kindred
spirits throughout, he has always made me laugh;
a lifelong friend before this miracle took place.
So easy when things are right to make a marriage last.
Who was to know we'd break the wedding news
of my third attempt? We surprised ourselves. I list

his attributes: he must be kind, with a love that lasts,
enjoy humour that makes us laugh, be full of news
and acceptance; live in a peaceful place. Everything ticked upon my list.

Red Moon of the Dhudhuroa People, January 2020

Campers watch crystal waters break
into sparkling swelling spumous waves,
curve around surfers straddling boards,
balance like dancers to stand upright,
bend and fold into huge tunnels that
cascade into thrashing frothy foam.

Tenders of the earth foam
at the mouth when faced with no break
from the dry, unbearable heat that
scalds the earth. Rain dancers wave
their arms willing rain to fall, right
where the parched earth is as hard as boards.

Lightning strikes ignite boards
and burn houses, start bush fires that foam
into flames that firefighters hose, working right
through the night. They create firebreaks
but the wind sends white waves
of ash far ahead, lighting outbreaks that

cover acres. Precious lakes provide water that
is utilised to quench the boards
of nearby towns that lie at risk of depraved waves
of erupting orange streaks of fiery foam
like clouds of molten anger, before evacuees flee in breaks
before roads are blocked, safety first is right.

Pets and children huddle on beaches right
before their once green campsite, the place that
they shared with friends; their holiday break
destroyed, now charred and smouldering boards,
with thoughts of sitting in the ocean foam
surrounded by the charcoal-embered waves.

Days and nights of frightening red waves,
smoke filled air, a molten sun isn't right.
Dream of home; not knowing if it's there. Foam
filled pillow, substitute for a lost teddy bear that
may be burnt. Waiting, waiting, no thought of being bored,
just waiting for the naval ship and rescue break.

Waves from sailors bring hope, a getaway from that
raging firestorm. Turn right to safely board
and watch the orange-tinged foam that the ship's wake breaks.

In the Time of Corona, March 2020

Neighbours wave and message, a change
from life's accelerated pace, where time
had spun out with no chance to exchange a kind
word. The carousel of life stopped, Corona
weaselled its way in and we changed course.
New hygiene methods must be taught.

Life as we knew it faltered. Folk taut
with anxiety; jobs lost, social change;
travel no longer a matter of course.
People re-evaluate what's important, time
is precious, long or short. Corona
brings uncertainty of a magnified kind.

An invidious virus of an unknown kind
is wildly contagious. Washing hands taught
multiple times. Wreaking havoc, Corona
jumps borders, bringing social change.
Public hoarding of toilet paper, selfish times
duplicate worldwide. Torment courses

through communities, where coarse
words and bun fights explode; yet kind
people surface sharing rosemary and thyme
and lift the spirits with home-cooked torte,
to bring equanimity and love in exchange
for the anger spreading from the Corona

Crisis. In Italy the pervading Corona
infects a country in lockdown, a course
taken too late. Social distancing is a change
inhibiting close social interaction, a kind
of isolation repugnant to most, where taut
laws restrict sport, learning and leisure time

activities. Isolation brings resilience in a time
of uncertainty for writers and artists. Corona
imposes time alone, time to create. Let taut
thoughts unwind and resolve in the course
of artistic endeavour. Solitude brings a kind
of perfection to a torfle world of change,

time of duality is part of the course.
Death by Corona is a shock of a kind
that is teaching patience and permanent change.

Francesca, 10.6.20

It's hard to believe fifty years have passed.
Several lives ago, you were a babe in arms.
A precious bundle with the eyes of a seer,
dark hair haloed your precious head.
'Che piedi nudi!' everyone said as we walked
in the summer sun, before we moved into isolation.

Witikon, Cabbio, a devoted time in isolation
where you thrived as the seasons passed.
Then Effretikon and Scheuren where we walked,
no longer a babe in arms.
A knowing beyond your years, thinking ahead.
'Do butterflies go rusty?' you asked, my quiet seer.

Always silent in Meeting, such a wise seer.
Reading, Ufton Nervet, another time of isolation
before Bert returned and Rebecca showed her head.
Friends speaking English, time passed
quickly with another babe in arms
who warmed our hearts as we walked.

Back to family in Devonport, again we walked,
grandparents delighted to share your seer-
like wisdom as you enjoyed their hugging arms.
No longer foreign, an outsider, or in isolation.
Loving years helped buffer what had passed.
You thrived in wider family, until you came head

to head with Ian. A step-dad, not an easy head-
space for you to accept. Hobart widely walked.
Brother Simon born before the year had passed.
Another year of student parents, life in Hobart seared
from us when Ian accepted work in Cairns. Isolation
yet again from friends and grandparents' arms.

Cairns sweltered, tears shed, school alms
exchanged for Freshwater where you became the head
of children who had been isolated
from their homes. You thrived with others, who walked
with you and gained trust in your seer-
like understanding as precious time passed.

From babe in arms, your life's like a swirl of smoke, you chose which path to walk.
Married James, two beautiful boys; moved to business where you became head seer.
Now through pandemic isolation, you thrive in work as fifty years have passed.

Rebecca, 10.4.21

Gentle as a raindrop, your spirit infuses
kindness to all within your sphere.
Your solicitude brings comfort
and your neighbour basks in care.
How wonderful for your children
to be nurtured with your love.

Rebecca, you were born in love,
and have returned it, to infuse
your family twofold. From a child,
you included others in your sphere.
In the group home your care
gave the less fortunate comfort.

I feel privileged for your comfort,
embraced by your supportive love.
Your awareness of the need to care
for animals and humans, infuses
me with pride. To bask in your sphere
brings joy and love for your children,

whom you have taught well. Your children
replicate your care and manifest comfort
to all of those within their sphere.
How salutary that you share love
and joy with dance and song, to infuse
the universe with your contented care.

Today you've found your niche for caring.
As the years pass by so quickly, your children
are growing strong. You have time to infuse
your work and bring your gifts of comfort
to those who seek your abilities and love
to beautify and imagine aesthetic spheres.

You recreate your dream homes in others' spheres,
and replenish your creative urges with resplendent care.
It's wonderful to see your expansive love
absorbed in work, Wade and the children;
bask in family life, filled with comfort,
as you surround yourselves in beauty that infuses

a joyous family sphere. Your growing children
share one another's care. You create comfort,
as the universe lights up, infusing you with love.

Simon

You incessantly asked to be born,
how could I resist? At two,
you loved picture book stories.
At five, your favourite thing was play.
That continues throughout your life,
from building cubbyhouses,

then transitioning to life-size houses.
You were patient to be borne
away; Fiona brought you new life,
a growing family, Finn and Reuben. Two
amazing boys give you an excuse to play
as they enrich your story.

Now you're building a second storey
onto your Melba house.
You strip it bare and play
with the plan for Fiona's plants, born
to flourish in light filled spaces. 'Too
many plants,' you say. 'Life

without plants is no life
at all!' So, you build on a second storey
to accommodate Fiona's two-
fold plan, to give space in the house
for her plants and your boys, borne
away with enthusiasm and with space to play.

No time to watch a play,
completeness of your lives
is replete. Fiona watches animals born;
as a vet she hears many stories.
You look after the children, house-
father of one where it used to be two.

Now it's time for you, too,
to complete another project. Play
became your career, building unique houses.
You have found your life's
purpose in helping others dream stories,
that evolve into homes where their imaginings are born.

So good to have a window between lockdowns to share your lives,
with a well-earned break here, to tree surf, play and read stories.
Your house now begun, we'll watch and wait, to see it reborn.

Crisis Brings Change, 20.9.20

Pandemic sweeps change;
businesses close, social disruption;
democracy turned on its head.
Poverty looms, livelihoods lost,
freedom a thing of the past;
isolation, desperation spreads.

Numbers escalate, Covid spreads.
Politicians fumble to change,
travel bans, curfews; dating now in the past.
Families trapped overseas, total disruption,
flights cancelled; people lost;
mental illness brought to a head.

Lockdown blankets the way ahead,
haunted uncertainty with viral spread.
Schools close, children feel lost;
elusive vaccination, longing for change.
Future ruination, wracked disruption.
Aged care homes watch the elderly pass.

We long for our golden past
while Zoom connects the way ahead.
Calm isolation, some feel disrupted.
Innovative musicians spread
collaborative music; bring positive change;
choirs emerge, heal those who feel lost.

Anguish masked faces, personalities lost,
hoarded toilet paper, no thing of the past.
Wash hands, keep distance, this hasn't changed.
Wear a mask, get tested, throat, nose or head-
colds; minor ailments, if ignored can spread
the virus; cause death and disruption.

Stay informed, avoid disruption.
Borders reopen, numbers lost.
Neighbourhoods share, kindness spreads
benevolence; vegetables, cakes pass
from one to another, before we head
home. All watch TV, willing a change.

The solution to disruption demands major change.
Critical thinking ahead, essential before our planet is lost.
Hope for vaccines vital to make viral spread a thing of the past.

Tormented Love

(after Audre Lord (1934–1992): 'Only by learning to live in harmony with your contradictions can you keep it all afloat.')

Do I ignore, forget, pretend or lie
about my contradictions or embrace
them? Who and why have I loved?
Complexity, simplicity, both encompass
inconsistencies that torment.
Love in truth brings harmony.

Stillness, clarity, harmony
beguile the sleeper as she lies
oblivious to past torment.
New beginnings embrace
further contradictions; but my compass
no longer swivels in search of love.

The gift of childhood love,
bestows a harmonious
endowment, a coat that encompasses
my being for a lifetime. A lie
can't cover the lack of an embrace.
To miss out on love is to live in torment

and self-doubt. Torture and torment
track an unworthiness to be loved.
Pity those with an embrace,
surround them in light and harmony
where forgiveness of lack of self-worth lies.
Love is to be shared. Encompass

all who've been deserted. A compass
for a new beginning where torment
is suffocated. Bring hope that doesn't lie.
Privilege is to love and be loved,
let contradictions dissolve in harmony.
Futures grapple in a fresh embrace.

Doubts dissolve in warm embrace.
Hope, an ever-ready compass
sets sail for an ocean of harmony
where tragedy and torment
are tipped overboard to drown in the depths of love.
Finally, at peace we lie.

Enfold into my embrace, torment
forgotten, encompass beloved
in songs of harmony where singing angels lie.

Furious Fiction

Words you've chosen inspire
me to write sestinas, a love of mine.
Words like rose, a flower or action; rose
in all of its glory, rosy-coloured or simply white,
bring other words to mind like light; imagine
feather weight or purely bright.

Three encouraging years have brought bright
writers together and have inspired
furious feasting on fast fiction. Imagine,
you have kept us focused, brains alert; mine
addicted to sestinas. White
pages turn to text, with a blush of rose

coloured harmony. A rose
that kept depression at bay. Bright
light, infectious good will; white,
black or in between, inspire
us to write flat out. Choose mine!
A lottery of so many to choose from. Imagine

furious fiction engrossing us all. Imagine
cheeks flushed with rose,
this time, will they choose mine?
Hundreds of writers, all bright,
fashion fiction to inspire.
Don't worry how bland or white,

ignore your uncertainties, white
out those doubts, just express, imagine.
Expose those hidden desires, inspire
others to write as you all rise.
Perfection isn't bright,
take imaginative risks like mine.

Writing is like drilling a mine,
to find sparkling diamonds so white
they dance with glittering bright
facets, blinding perspective, just imagine.
Your magnificent prompts, like a rose,
they unfurl. Starters pistol, go, go, go! inspire!

If you choose mine, I could only imagine,
I'd throw white petals to bless tenses forlorn. A rose
is my congratulations to those bright minds who inspire.

Ne'er Forgotten

Grandma's hug, soft and doughy,
holds me on her knee. Silent,
still, I watch a heart-shaped
shadow of the pear tree rock.
Sunset golden glow long passed,
I'd woken from a dream. Would

the stark sky turn to black? Would
there be a moon? Her doughy
warmth enables dreams to pass.
She tucks me in, a silent
kiss; no longer there to rock.
I watch her bent nightie shape

disappear as shadow shapes
grow tall. I tighten fists; the wooden
staircase creaks, a rock-a-bye
creak. Hidden; the pillow's doughy
softness suppresses silent
screams. I whimper, shadows pass.

Was that a ghost gliding past,
or am I in a dream? Shapes,
razor sharp, rake silently.
I close my eyes so I won't
see any weird or doughy
scary creatures fly or rock

by. Lying flat on aged rock,
blue and yellow fish swim past,
soft sponges spring with doughy
touch to make a pillow shape
to rest my head. A wooden
sail-boat glides by silently.

I listen to the silence,
happy on this sun warmed rock.
I watch a flock of wood pigeons
gather in formation, pass
beyond the shoreline shape
of nether worlds so doughy.

Gran's delicate perfume silently wafts as she goes past,
I rock into wakefulness as I hear her words gently shape,
'Would you like jam as thick as honey on your doughy bread?'

A synopsis of men who passed through and the one who stayed

Rape, too harsh a word,
though ready, I was not.
Lust pumped through my blood.
My mother's red kilt wore my guilt.
Old George, the potter, lifted off
'Oh, you were a virgin,' is all he said.

I painted red until Umberto said,
'You're mine, you gave your word.'
Fate had plans for us to travel off
to Italy after marriage and question not.
Eight years, two daughters later, he felt no guilt
and left us in the UK. We returned to my blood

relatives, to protect us from his tainted blood.
Supportive family, my mother said,
'Don't hold bitterness inside.' Released to gilt
edged widower, Peter, dearly loved. His words
resonated, I flourished, though stepmother I was not.
I left to study and took my girls off

to Hobart where I fell immediately off
the wagon, parsley germinated, blood
vanished. To Quaker Ian, I fell in love, not
looking past my nose. He hadn't said
he was unemployed; we married. Harsh words
from home left me feeling guilt.

We travelled three states; ten years on, guilt
stripped from my younger self. Ian transferred, off
again, a yen always for the new. I stayed put with word
of permanent work; the children happy, no blood
shed. John came to help prune trees, Beck said,
'He's so nice, I'll take him a cup of tea.' Not

the last of those he had. A decade later, not
too late to dream, another missive, gilt
edged, changed my life. Christopher said,
'Come visit me for the opening,' so off
I flew. Chris changed my life, my blood
pounded once again, when he pressed the words,

'Marry me!' With not a second thought, I knew. Off
without guilt; our children grown, no blood
spilt, bound as was said in the fortune-teller's words.

Quakers

Let a seed drop on the ground
in cracked pavement, roots expand;
like a prayer in meeting allows the silence
to work its magic. Pacifist folk
supported the young men who refused
to go to the Vietnam War. Refreshing

to meet people unafraid, refreshing
to hear they don't know. Ground
rules changed; no dogma; they refuse
to follow traditional lines. Expanded
minds, open to new ideas. Folk
find themselves in spiritual silence.

Attending a meeting, in the silence
an elder spoke to my condition. Refreshed
to find telepathy works, spiritual folk,
humble, unworldly, yet grounded.
I'd found my kin. Supported to expand
my constraints, never to be refused.

No proselytising. No refusal
when Umberto and I asked to join. Silently
we became Quakers, ready to expand
our horizons. Refreshed,
ready to meet other grounded
and like-minded folk.

United thoughts, no matter folks'
language, combine, never refuse
to help understand and find united ground.
Inspiration abounds in silence.
People moved, speak refreshing
ideas that allow the mind to expand.

Small groups of Quakers expand
around the world, uniting folk
to improve the planet; refreshing
purpose for wells where water is refused.
When united in silence,
arguments go to ground.

I no longer attend, still prayers expand. United in refusal
of the status quo. Support voiceless folk; join in silent
refreshing movement to protect this, our sacred ground.

Ablutions

Palpating strokes soothe my skin
from a water-saving shower head.
Like a lily in the rain
I receive comfort from the stream.
Ideas procreate and disappear
before I can retain them, they flit,

dive like a dragonfly on tequila. Flit
away to another world, a skin-
like film suppresses; then disappears.
Retention is a skill to learn, my head
is flawed, where thoughts stream
out as well as in. Dreams reign

in the night, the gift of capture reigns
by day. I soap myself and let the flit
of foam lather. Let other thoughts stream
through my mind while soaping skin,
thoughts buzz and tickle inside my head.
Conditioned? Short-term memory disappears.

I repeat in case and rinse my disappearing
thoughts down the drain. My love of rain
is nurtured, the gentle thrumming shower-head
pulses; relaxes; letting past voices flit,
to remind me of their wisdom. Slippery skin,
softer now, released from streaming

tension. I hear voices streaming,
Nana first, 'dry between your toes,' then disappears.
Mary and June remind me to moisturise my skin,
and 'OO EE OO EE, to tighten under chin.' Rain
patters with a laugh, 'it doesn't work,' then flits
away. 'I know that,' I say inside my head,

'but, worth a try.' I think of my day ahead,
before turning off the streaming.
I dry with a soft white towel and flit
a glance at senile warts and wish they'd disappear.
So grateful to hear the rain
outside, it makes the thought of skinny

dipping froth inside my head before it disappears.
The days of swimming in the stream as soaking rain
pelts down; such fleeting memories tweak my skin.

Quiddity 2021

I feel like someone getting old;
recognise my identity is slipping
through the gap of fence, where
birds sit still and sing. I watch
as people strive to change
the rules, the cost of living rises.

The cost of houses rises;
Baby Boomers blamed. Old
houses remain unchanged
when old folk stay until they slip
and have a fall. They watch
with horror when they're put where

no one wants to be. That's where,
in care homes, the panic rises.
No one wants to see others die and watch
rich managers spend funds for the old
turn profits for investors. Slip
food standards for loose change

to keep investors happy. Such change
is inappropriate. Federal funding is where
money for the boys is obnoxiously slipping.
Care home. The name belies the rise
of cutting staff and foodstuffs for the old,
as was deemed in the Royal Commission. Watch

and wait while nothing's done. Watch
frightened as people become disabled, no change
for those with dementia, those old,
unable to speak themselves, this is where
respect and dignity must be improved. A rise
of demonstrations from our young before they slip

themselves into the trap of poverty. A slip
of paper regarding measures to politicians, we watch
for change before the next election. If people rise
together; tell them what we think, a change
is imminent, if we all speak up together. Where
are the voices of the young, mixed with the old?

I pray in this sestina to not let us slip and suffer change.
Chris and I shall keep watch; let's hope we can stay home, where
we're able to rise to write and paint, enjoy life fully until we're really old.

If

It takes one fall to change one's world.
Lost; twin soul dependence.
Severance extinguishes light,
gone essential treasured piece,
the complex jigsaw no longer feels right
now you have passed. You left

me alone; your loss is immense. Left
to survive a world
where no one can replace you, right-
fully mine. You gave me independence,
shared wonders. No longer at peace,
dark silence eclipses light.

I miss your chatter, music and light,
pretend to be brave; but you left
shards in my heart, pierced pieces.
Loved ones sustain my world,
weft through my warp. Independence
taught when I was a child; sent right

after primary school to board; right
felt wrong. Alone, amongst the hordes, the grey light
of the dormitory, I learnt, like you, independence.
Indelible memories of you will never leave
me; my first ball, in that new world
with you, brought me a sense of peace.

Farm holidays, you and Clive, a piece
of history that felt right.
My brother's best friend; we shared a world
where our dreams were entwined. The light
of your enthusiasm encouraged me to leave
school and follow you to real independence

at art school. There, you showed me independence,
took me to lunch to decide which piece
of the puzzle would fit, chose my mate, and later left
me your job, when you went overseas. It was so right.
I had a year at home with my family, engulfed in the light.
So lucky, before stepping into a foreign world,

where confusion didn't deplete my rightful independence.
Now, your peaceful spirit surrounds me with your light.
If you leave, you told me, you'd remain by my side in this world.

Opposites Matter

Valued thoughts lie deep inside
like bowerbirds lining nests
with flowers, berries and pegs of blue,
keeping treasures safe,
as outsiders trample, blind,
oblivious to a cherished gift.

Screened by aegis, my hidden gift
kept pillowed, warm and soft inside.
Shredded comments scatter blind
like scarlet flint, beyond the nest
where secrets snuggle safe
flushed with ultramarine, my favourite blue.

Songs in depths of perfect blue
that render memories are a gift.
Battered and bruised, now stored safe
like data, are transposed inside
a cocooned fluffy nest
where life's dross can hide. Blind

fools can't see, they are blind
to what lies concealed. Cerulean blue
straws protect and bind this nest,
entwined to cosset this fragile gift.
Cerulean colour calms inside,
transforms to make it safe.

The mind's harvest lies safe,
where angels delve. Blinded,
I rummage to retrieve; inside
imbued with glorious blues;
Prussian, cobalt, indigo, gifts
that entwine and cosset, nestle

stranding this precious nest.
Ancient mishaps stored safe,
paper trails burnt, my gift
remains strong within. How blind
when I cannot see the blue
that heals my psyche and cleanses me inside.

Lapis lazuli builds strength, the nest has no blind
to shadow. I unlock the safe within, immerse my soul in the purest blue,
to reveal the gift of self-love that radiates golden light from inside.

Covid 19 and the Arts (double sestina)

Covid 19 generates momentous change.
Lockdown enforces a new way of life.
The Arts infiltrate essence in so many ways,
reading or watching TV. Absorption leaves
isolation and boredom at bay,
time enhanced fundamentally with the Arts.

Cultural dancing, or learning the art
of tap or ballet to implement change.
Jazz moves or cancan will keep sloth at bay.
Waltz with a broom if alone in your life,
or watch Fred Astaire if you can't leave
your chair. Just twist and turn; dance many ways.

Oils or watercolour paint are two ways,
acrylic, draw or make collage, art's
not restricted to paper and canvas; walls leave
space to manifest change.
Enhance children's bedrooms, life's
colour; bring enchantment to bay.

Isolation; we all like to eat. Pick some bay
leaves and add to the soup; always
share with the neighbour who lives
alone. Cooking, an ancient and modern art
form brings fresh new ingredients for change.
Healthy eating includes lots of red and green leaves.

Musicians have come to the fore without leaving
their homes. Keep insanity at bay
by Zooming choirs and YouTube renditions of change
that excite children and adults alike, sharing ways
of enjoyment and joyous fun. Using musical skills, art
uplifts spirit to improve family life.

Lockdown, perfection for a writer's life,
means jobs for journalists, forced to leave.
Unemployment has been harsh on the arts.
Those needing to publish, bay
at the Murdoch moon; write poetry in ways,
accelerate acceptance for great need for change.

The performing arts: opera, stage or film await change.
Performances cancelled, peoples' lives
on hold. Composers continue in new ways,
have overseas bookings, hope to leave
and not stay in lockdown; feeling like dogs baying,
howling, as footballers play. Their neglected art

forgotten. Designers of furniture and fashion artfully
set up tables and continue with changes
to work from home; whilst children, with computers, bay
at their classmates on screen. Schools are closed, so life
is trapped within four walls. Pity for some who left
their homes; stuck, forced to stay away.

Lucky the ones who spin or weave; the ways
and means for keeping calm with a gentle art.
Time lapses for those so involved that leaves
them wondering the day of the week, as change
happens silently for the privileged few. Life
in isolation keeps the virus at bay.

Seek vaccination when vaccines are scarce, bay
all you like, patience required. Crafty ways
like ceramics, building pots, bring to life
skills on the wheel. A true artisan
yields pleasure from creations that change
form into inspirational beauty that leaves

folk wanting more. Garden designers leave
growing landscapes renewed with Bay
hedges for generations to come. Changeable
fashion is not something new; finding ways
making habitat for animals and birds; artfully
create spaces that transform peoples' lives.

Architecture provides solace to lives,
whether semi, or terraced, our abode leaves
us with respite. During this lockdown no art
was spared from offspring keeping problems at bay.
Mental illness recurrent with returns to the nest, ways
hauntingly common; parents seek options for change.

Chris and I feel so fortunate, life continues, keeping at bay
the problems we see. We're left to paint and write in our own way,
allowing the arts to enhance our lives, observing the change.

Sinuous Sestina Spiral

The tendrils of a sestina
slowly cling, curl and criss-cross,
weasel their way
to coil like a Nautilus shell,
making a pyramid spiral
until the lines tie with a knot.

Addiction holds fast if you're not
alert; to slither into yet another sestina
before you can spiral
around to begin a crossword.
Another story from an empty shell
starts before it wriggles away,

to find its own way
to divulge secrets that I've promised not
to tell; they just squirm out of the shell.
If a page gets torn, the sestina's
sinuous corkscrew gets cross,
its backbone is broken, the spiral

unfurls or is torn. A patched spiral
is never the same. Words are so way-
ward and contrite, best not to cross.
Patch paper together and hope that words knot
to reform into the pattern a serpent sestina
commands; rehabilitate back to its shell.

Reimagine another shape, the shell
of a church, with rafters spiralling
into a vault. A cathedral sestina;
stair winding its own glorious way
to a slated dome forming the topknot.
Immutable atop is a shiny gold cross.

Words become family that, at times, make me cross.
Old age brings awareness of my vulnerable shell.
Lockdown, I stay in my dressing gown, knotted;
able to scribble or type words that spiral
into poems, those tortuous throw away
thoughts that form a compendium of sestinas.

For those I have crossed, my thoughts spiral
anew; forgive me before I crawl into my shell; to weigh,
contemplate, a Covid *envoi* to complete this knotted sestina.